Mangrove Honey

Unlocking the Benefits of Beekeeping for Coastal Ecosystems and Communities

Sadiq A. Al Qatari

Index

1. Introduction to Mangrove Honey
2. The Role of Mangroves in Coastal Ecosystems
3. Beekeeping in Mangrove Habitats: An Overview
4. The Science Behind Mangrove Honey Production
5. Economic Benefits for Coastal Communities
6. Sustainable Beekeeping Practices and Techniques
7. Environmental Impacts and Ecological Benefits
8. Case Studies: Successful Mangrove Beekeeping Projects
9. Challenges and Solutions in Mangrove Beekeeping
10. Future Prospects for Mangrove Honey and Conservation
11. Policy and Support for Mangrove Conservation and Beekeeping
12. Conclusion: The Sweet Path to Sustainability

Introduction to Mangrove Honey

Mangrove honey, often referred to as "liquid gold," is more than just a sweet treat; it embodies a unique convergence of nature, culture, and sustainability. Derived from bees that forage in the blossoms of mangrove forests, this honey captures the essence of coastal ecosystems. Its rich flavor profile, distinct from conventional honeys, reflects the briny air, lush foliage, and floral notes unique to mangrove environments. However, the significance of mangrove honey goes far beyond its taste. It is deeply intertwined with ecological preservation, sustainable livelihoods, and community resilience, making it a fascinating subject in the realm of natural resource management.

The production of mangrove honey offers numerous environmental, economic, and cultural benefits. By understanding the intricate relationship between beekeeping and mangrove ecosystems, we can better appreciate how this sweet product contributes to coastal sustainability and the well-being of communities who live near these coastal habitats. As interest in sustainable beekeeping continues to grow, mangrove honey presents a compelling case for the intersection of traditional practices and modern environmentalism, providing a path toward economic empowerment and ecosystem conservation.

The Unique Characteristics of Mangrove Honey

Mangrove honey possesses a distinct character, thanks to the complex floral sources available in coastal forests. Mangroves, known for their resilience in saline environments, support a wide array of flowering species, including red, black, and white mangrove varieties. The nectar from these blossoms infuses the honey with a unique combination of flavors—often described as robust, with earthy and slightly smoky undertones. This honey also tends to be darker in color compared to traditional floral honeys, with a thicker consistency and higher mineral content, which contributes to its growing popularity among honey connoisseurs.

The bees that produce mangrove honey play a crucial role in maintaining the health of these coastal ecosystems. As they collect nectar, they facilitate the pollination of mangrove trees and other plants in the area, ensuring the continued growth and resilience of the forest. This natural pollination process not only sustains honey production but also supports the broader ecological functions of mangrove forests, such as carbon sequestration, shoreline stabilization, and providing habitats for diverse wildlife species.

The Ecological Importance of Mangroves

Mangrove forests are often described as "blue forests" due to their capacity to act as carbon sinks, trapping atmospheric carbon dioxide more efficiently than many terrestrial forests. They play a vital role in mitigating the effects of climate change by sequestering significant amounts of carbon, a process that is amplified when mangroves are well-maintained and continuously pollinated by bees. Additionally, the intricate root systems of mangroves help protect coastlines from erosion, storm surges, and tsunamis, creating a natural barrier that shields inland areas from environmental disturbances.

These forests also serve as nurseries for numerous marine species, such as fish, crabs, and shrimp, which rely on the shelter provided by the dense mangrove roots during the early stages of their life cycles. By supporting a thriving bee population within these habitats, mangrove honey production helps sustain the ecological balance that is crucial for the survival of countless organisms. This delicate web of life illustrates how the simple act of beekeeping can have a profound impact on broader environmental health.

Economic and Social Impacts of Mangrove Honey Production

For many coastal communities, beekeeping in mangrove areas offers an alternative source of income that complements traditional livelihoods like fishing and agriculture. In regions where these primary occupations are becoming increasingly unpredictable due to climate change and overfishing, the cultivation of mangrove honey presents a sustainable means of economic diversification. The sale of high-quality, organic mangrove honey, often at premium prices in local and international markets, allows communities to generate additional income while contributing to the preservation of their natural environment.

Training in beekeeping also provides opportunities for skill development, particularly for women and young people who might otherwise have limited employment prospects. By learning the craft of beekeeping, they can not only earn an income but also contribute to community resilience in the face of economic challenges. Beekeeping programs in coastal regions often incorporate educational components that raise awareness about the importance of mangrove ecosystems and the need for sustainable environmental practices, thereby fostering a sense of stewardship among local populations.

Cultural Significance and Traditional Practices

The practice of harvesting honey from coastal forests is not new. Many cultures have long valued honey not only as a food source but also for its medicinal and spiritual properties. In some coastal societies, traditional beekeeping techniques have been passed down through generations, with local knowledge and skills adapted to the unique conditions of mangrove environments. These practices often emphasize sustainable methods that minimize disturbance to the bees and the surrounding habitat, aligning with modern principles of natural beekeeping.

In places like Southeast Asia, Africa, and the Caribbean, mangrove honey is prized for its therapeutic qualities, believed to possess antibacterial, anti-inflammatory, and antioxidant properties. Local customs often include the use of this honey in traditional medicine, rituals, and cultural festivals, where it symbolizes the sweetness of life and the interconnectedness of nature and community. By integrating traditional knowledge with contemporary sustainable practices, beekeeping in mangrove regions serves as a bridge between cultural heritage and modern environmentalism.

Challenges in Mangrove Honey Production

Despite its numerous benefits, the production of mangrove honey faces several challenges. Coastal areas are often under pressure from development, pollution, and deforestation, which threaten the health of mangrove forests. When these ecosystems are compromised, the availability of nectar sources for bees diminishes, leading to lower honey yields and potentially weakening bee colonies. Additionally, climate change can disrupt the flowering cycles of mangroves and other coastal plants, further affecting honey production.

Another challenge lies in balancing the economic incentives for mangrove honey production with the need for environmental protection. While the expansion of beekeeping can support mangrove conservation, it is crucial to ensure that it does not lead to the overexploitation of natural resources. Sustainable practices, such as limiting the number of hives per area and adhering to ethical harvesting methods, must be prioritized to protect the long-term health of the ecosystem.

The Path Forward: Opportunities for Sustainable Growth

To fully realize the potential of mangrove honey as a tool for conservation and economic development, it is essential to adopt a holistic approach that integrates beekeeping with broader environmental and community initiatives. Efforts to restore and protect mangrove forests should go hand-in-hand with the promotion of sustainable beekeeping practices, ensuring that honey production remains ecologically and economically viable.

Investing in research to improve beekeeping techniques, understanding the specific nectar sources in mangrove ecosystems, and developing better marketing strategies for mangrove honey can help increase the value of this unique product. Furthermore, partnerships between governments, non-profit organizations, and local communities can provide the necessary resources and support to expand sustainable mangrove honey production while protecting coastal ecosystems.

Mangrove honey represents more than just a product, it is a testament to the harmonious relationship between humans and nature. By tapping into the potential of beekeeping in mangrove forests, we not only produce a valuable natural resource but also contribute to the preservation of some of the planet's most vital ecosystems. As awareness of the ecological, economic, and cultural benefits of mangrove honey continues to grow, so too does the opportunity to use this "liquid gold" as a catalyst for sustainable development and environmental stewardship.

The Role of Mangroves in Coastal Ecosystems

Mangrove forests, sometimes referred to as the "guardians of the coast," are a vital component of coastal ecosystems around the world. These unique trees and shrubs thrive in the intertidal zones of tropical and subtropical regions, where they act as a crucial buffer between the land and sea. The intricate root systems of mangroves allow them to withstand saline waters and tidal fluctuations, while also providing significant ecological, economic, and social benefits. Despite their importance, these ecosystems face numerous threats, making their protection and restoration more critical than ever.

Ecological Importance of Mangroves

Mangrove ecosystems play a fundamental role in maintaining ecological balance along coastlines. One of their most essential functions is providing a habitat for diverse marine and terrestrial species. The tangled root structures create sheltered nurseries for fish, crustaceans, and mollusks, which find refuge from predators and harsh environmental conditions. Many commercially valuable species, such as shrimp, crabs, and fish, rely on mangrove forests during various stages of their life cycles. In this way, mangroves not only sustain biodiversity but also support local fisheries and economies.

In addition to marine life, mangrove ecosystems host a variety of bird species, reptiles, and even mammals that depend on this habitat for feeding, nesting, and breeding. Migratory birds, for example, use mangrove forests as stopover sites to rest and refuel during long journeys. By supporting a wide range of species, mangroves contribute to the overall health of coastal biodiversity, enhancing the resilience of ecosystems against disturbances.

Coastal Protection and Erosion Control

One of the most remarkable attributes of mangrove forests is their ability to protect coastlines from erosion, storm surges, and tsunamis. The dense network of roots helps trap sediments and stabilize the shoreline, preventing soil loss and reducing the impact of wave action. During extreme weather events, such as hurricanes or cyclones, mangroves act as a natural barrier, dissipating wave energy and reducing the force with which these waves reach the shore. This buffering effect significantly lowers the risk of flooding and property damage, offering protection to human settlements, agricultural land, and infrastructure located in coastal areas.

The roots' ability to trap sediments also helps to maintain water quality by filtering out pollutants and excess nutrients that may otherwise flow into coastal waters. This filtration process prevents harmful substances from reaching coral reefs and seagrass beds, which are sensitive to changes in water quality. By enhancing coastal water clarity, mangroves indirectly support the health of adjacent marine ecosystems, contributing to the overall productivity of the coastal zone.

Carbon Sequestration and Climate Regulation

Mangrove forests are often considered one of the most effective natural solutions for climate change mitigation, thanks to their significant carbon sequestration capacity. Mangroves are capable of capturing and storing large amounts of carbon dioxide (CO_2) from the atmosphere, a process known as "blue carbon" storage. Their ability to sequester carbon is particularly impressive because they not only store carbon in their biomass (leaves, stems, and roots) but also in the rich organic sediments that accumulate in the mud beneath them .

Studies show that mangroves can sequester up to five times more carbon per hectare than terrestrial forests, making them a crucial ally in the fight against global warming. When mangrove forests are degraded or destroyed, however, the stored carbon is released back into the atmosphere, contributing to greenhouse gas emissions. Therefore, protecting and restoring mangrove forests is a valuable strategy for climate regulation and reducing the impacts of climate change.

Support for Livelihoods and Local Economies

Beyond their ecological benefits, mangrove ecosystems play a significant role in supporting the livelihoods of millions of people who live in coastal areas. Many communities rely on the resources provided by mangroves for food, medicine, and building materials. The forests supply timber and non-timber products such as firewood, thatch, and tannins used in traditional medicine. Additionally, the abundance of fish and shellfish within mangrove areas is a critical source of protein for local populations.

Mangrove-based ecotourism also presents an opportunity for economic development in coastal regions. Activities like birdwatching, kayaking, and nature tours attract tourists who seek to experience the natural beauty and biodiversity of mangrove forests. When managed sustainably, ecotourism can generate income for local communities while raising awareness about the importance of mangrove conservation. Training locals to act as guides and conservation advocates not only creates jobs but also fosters a sense of pride and ownership in the preservation of these critical ecosystems.

Threats to Mangrove Ecosystems

Despite their importance, mangrove forests face several significant threats. Coastal development, including the construction of ports, resorts, and aquaculture facilities, often leads to the clearing of mangrove areas. Land reclamation for agriculture and urban expansion also contributes to habitat loss, as mangroves are removed to make way for other land uses. This loss of habitat not only disrupts the ecological functions of mangrove forests but also undermines their role in protecting coastlines and supporting livelihoods.

Pollution poses another serious threat to mangroves. Oil spills, agricultural runoff, and untreated wastewater can introduce toxic substances into mangrove environments, damaging plant life and reducing the forests' ability to filter pollutants. Additionally, changes in freshwater flow due to dam construction or irrigation projects can alter the salinity levels in mangrove ecosystems, making it difficult for these salt-tolerant species to thrive.

Climate change further exacerbates the challenges faced by mangroves. Rising sea levels can submerge low-lying mangrove areas, while increased temperatures and shifting weather patterns can affect the timing and intensity of flowering and seed dispersal. These changes can impact the reproductive success and overall resilience of mangrove forests, making them more vulnerable to other stressors.

Mangrove Restoration and Conservation Efforts

In response to the growing threats to mangrove ecosystems, various conservation and restoration initiatives have been launched worldwide. Restoration projects typically involve the replanting of mangrove species in degraded areas, with the goal of re-establishing natural ecological processes. Successful restoration requires an understanding of local conditions, such as soil type, tidal patterns, and salinity levels, to ensure that the selected mangrove species are well-suited to the environment.

Community involvement is often a key component of mangrove restoration efforts. By engaging local populations in planting and monitoring activities, conservation projects can foster a sense of ownership and stewardship. This participatory approach not only enhances the chances of success for restoration projects but also ensures that local needs and knowledge are incorporated into management strategies.

Innovative approaches to mangrove conservation are also emerging, such as the use of blue carbon credits to incentivize the protection of these ecosystems. By quantifying the carbon storage benefits of mangrove forests, conservation organizations can sell carbon credits to companies seeking to offset their emissions. The revenue generated from these credits can then be reinvested in conservation activities, creating a sustainable funding model for mangrove protection.

The Global Importance of Mangrove Ecosystems

The significance of mangrove forests extends far beyond the coastal regions where they grow. These ecosystems play a crucial role in global biodiversity, climate regulation, and human well-being. As we continue to face the challenges of climate change, habitat loss, and environmental degradation, the protection and restoration of mangrove forests are more important than ever .

Governments, non-governmental organizations, and local communities must work together to implement policies and practices that prioritize the conservation of mangroves. This collaboration can include establishing protected areas, regulating coastal development, and promoting sustainable resource use. With concerted efforts, it is possible to halt the decline of mangrove ecosystems and secure their benefits for future generations.

Mangrove forests are a cornerstone of coastal resilience, providing essential services that protect shorelines, support biodiversity, and sustain human livelihoods. Their ability to sequester carbon and buffer against environmental disturbances underscores their critical role in addressing some of the world's most pressing environmental challenges. Preserving these unique ecosystems requires a holistic approach that recognizes their ecological, economic, and cultural value. By championing the conservation and restoration of mangroves, we can ensure a sustainable and prosperous future for both people and the planet.

Beekeeping in Mangrove Habitats: An Overview

Beekeeping, or apiculture, has long been practiced worldwide, providing communities with honey, beeswax, and other products while supporting agricultural pollination. However, beekeeping in mangrove habitats is a relatively specialized niche that brings unique benefits and challenges. Mangrove ecosystems—characterized by salt-tolerant trees and shrubs that grow along tropical and subtropical coastlines—create a distinct environment for bees to thrive. The resulting mangrove honey, known for its unique taste and medicinal properties, has grown in popularity as consumers become more interested in natural and sustainable products.

This article will explore the significance of beekeeping in mangrove habitats, the specific benefits of mangrove honey, the ecological contributions of bees within these ecosystems, and the challenges and opportunities involved in this specialized form of apiculture. Through a deeper understanding of these aspects, we can better appreciate the role that mangrove beekeeping plays in sustainable livelihoods, conservation efforts, and the broader environmental context.

The Characteristics of Mangrove Ecosystems

Mangroves occupy coastal areas where land meets the sea, creating a dynamic environment characterized by brackish water, tidal fluctuations, and muddy soils. These conditions give rise to an ecosystem with unique biodiversity and a variety of specialized plant and animal species. Mangrove trees have adapted to these challenging conditions with specialized root systems that allow them to survive in saline water and stabilize the shoreline against erosion. Mangrove forests also support a complex web of life, providing habitat and nursery grounds for fish, crustaceans, birds, and other wildlife.

The distinct floral characteristics of mangrove ecosystems make them suitable for beekeeping. The flowers of mangrove trees, such as the red, black, and white mangroves, produce nectar that is collected by bees to make honey. Because the trees blossom at different times of the year, mangrove habitats can offer a relatively continuous supply of nectar, making them an attractive environment for apiculture. In addition to providing a steady food source for bees, mangroves also play a crucial role in maintaining the health of coastal ecosystems by sequestering carbon, filtering pollutants, and protecting shorelines.

The Unique Qualities of Mangrove Honey

Mangrove honey is known for its distinctive taste, color, and nutritional profile. It tends to be darker than most conventional honey, with a rich, earthy flavor often accompanied by slightly salty, smoky, or caramel-like undertones. This unique flavor profile results from the specific nectar sources in mangrove environments, where the briny air and complex floral composition influence the honey's taste. Mangrove honey is also typically high in minerals and antioxidants, owing to the nutrient-rich environment in which it is produced.

Beyond its culinary appeal, mangrove honey is valued for its medicinal properties. It is often used in traditional medicine to treat various ailments, such as sore throats, digestive issues, and skin infections, due to its antibacterial, anti-inflammatory, and antioxidant qualities. The health benefits of mangrove honey have contributed to its increasing demand in local and international markets, with consumers seeking natural and organic products that also support sustainable practices.

Ecological Contributions of Beekeeping in Mangrove Habitats

Beekeeping in mangrove habitats does more than produce honey—it also supports the health and resilience of the ecosystem itself. Bees play a vital role in pollinating mangrove trees and other coastal plants, facilitating their reproduction and growth. This natural pollination process helps sustain the mangrove forest's vitality, promoting the continuous regeneration of trees that protect the shoreline and provide habitat for various species.

Pollination is especially important for mangrove forests that face natural and human-induced pressures, such as coastal development, pollution, and climate change. As bees contribute to the stability and regeneration of these forests, beekeeping can be seen as a complementary conservation activity that benefits the environment. The presence of managed bee colonies in mangrove areas can enhance local biodiversity and improve the resilience of the ecosystem to disturbances.

In addition, bees are known to pollinate other flowering plants that grow near mangrove forests, such as coconut palms and fruit trees. This cross-pollination can increase agricultural productivity in surrounding areas, providing added economic benefits to local communities. Thus, the ecological impact of beekeeping extends beyond the immediate environment of the mangrove forest, influencing adjacent agricultural and natural ecosystems.

Economic and Social Benefits of Mangrove Beekeeping

Beekeeping in mangrove habitats offers significant economic opportunities for coastal communities. The sale of mangrove honey and other bee products, such as beeswax, propolis, and royal jelly, can provide a sustainable source of income. This is particularly important in regions where traditional livelihoods, such as fishing and agriculture, are threatened by overexploitation, climate change, or economic instability. By diversifying income sources, communities can become more resilient to economic shocks and environmental changes.

The practice of beekeeping also promotes skill development and knowledge sharing, especially among marginalized groups, such as women and young people. Training in apiculture can empower these groups by providing them with the skills and resources needed to generate income independently. Moreover, beekeeping initiatives often involve educational programs that raise awareness about the importance of mangrove conservation and sustainable environmental practices. This creates a culture of stewardship in which communities actively participate in protecting their natural resources.

In addition to generating direct income from honey sales, beekeeping in mangrove areas can support the growth of ecotourism. Visitors who are interested in experiencing the natural beauty and biodiversity of mangrove forests may be attracted to activities such as honey tasting, guided tours of apiaries, and educational workshops on sustainable beekeeping. This form of tourism not only generates revenue but also raises awareness about the conservation of mangrove ecosystems, promoting a more sustainable approach to tourism development.

Challenges of Beekeeping in Mangrove Habitats

While the benefits of beekeeping in mangrove habitats are significant, there are also several challenges that must be addressed to ensure the sustainability of this practice. One of the primary challenges is the vulnerability of mangrove forests to environmental degradation. Coastal development, pollution, and deforestation pose major threats to these ecosystems, potentially reducing the availability of nectar sources for bees and leading to lower honey yields. When mangrove forests are destroyed, the natural balance is disrupted, and the benefits associated with beekeeping, such as pollination, biodiversity support, and coastal protection, are diminished.

Another challenge is the potential impact of climate change on mangrove habitats. Rising sea levels, increased temperatures, and changing weather patterns can affect the growth and health of mangrove trees, altering the availability and timing of nectar production. For example, prolonged droughts or higher salinity levels due to reduced freshwater inflows can stress mangrove trees, reducing their flowering and subsequent honey production. Beekeepers must adapt to these changing conditions by implementing strategies such as relocating hives, diversifying forage sources, or adopting drought-resistant bee species.

Managing bee health is also a critical consideration in mangrove beekeeping. While mangrove ecosystems provide natural foraging grounds for bees, they can still be affected by diseases, pests, and pesticide exposure. Monitoring bee colonies for signs of disease, maintaining hive hygiene, and using integrated pest management practices are essential to ensure the health and productivity of the bees. Additionally, beekeepers should be cautious about introducing non-native bee species that could outcompete local pollinators or spread diseases.

Strategies for Sustainable Mangrove Beekeeping

To maximize the benefits and overcome the challenges associated with mangrove beekeeping, it is important to adopt sustainable practices that prioritize environmental conservation and community well-being. Here are some strategies for ensuring the sustainability of mangrove beekeeping:

1. **Protecting and Restoring Mangrove Forests**

   The foundation of successful mangrove beekeeping lies in the conservation of mangrove habitats. Efforts to protect existing mangrove forests from deforestation, pollution, and development are crucial for maintaining the availability of nectar sources for bees. Restoration initiatives that involve replanting mangrove trees in degraded areas can also help regenerate ecosystems, improve biodiversity, and increase the resilience of the habitat.

2. Integrating Beekeeping with Conservation Efforts

Beekeeping projects should be designed in a way that complements broader conservation goals. This can involve collaborating with local environmental organizations, government agencies, and communities to integrate beekeeping into mangrove protection programs. By aligning apiculture with conservation efforts, it is possible to create a mutually beneficial relationship in which beekeeping supports ecosystem health and conservation supports beekeeping success.

3. Training and Capacity Building for Local Communities

Providing training in sustainable beekeeping practices is essential for building the capacity of local communities to manage beekeeping enterprises effectively. Training programs should cover topics such as hive management, disease prevention, honey harvesting, and marketing. Capacity-building efforts should also include education on the ecological importance of mangroves and the role of bees in maintaining healthy ecosystems.

4. Diversifying Forage Sources and Bee Products

To mitigate the risks associated with fluctuations in nectar availability, beekeepers can diversify the forage sources for their bees by planting additional flowering plants near mangrove areas. This approach can help ensure a steady supply of nectar throughout the year and improve colony health. Beekeepers can also diversify their products by exploring value-added goods, such as beeswax candles, propolis tinctures, and honey-based skincare products, which can increase revenue streams.

5. Promoting Sustainable Marketing and Certification

Marketing mangrove honey as a premium, organic, and sustainably produced product can help increase its market value and create incentives for conservation. Certification programs, such as organic or fair-trade labels, can further enhance consumer trust and support ethical practices. By promoting the unique qualities of mangrove honey and its ecological benefits, beekeepers can tap into niche markets that value sustainability and natural products.

Case Studies: Successful Mangrove Beekeeping Projects

Several successful mangrove beekeeping projects around the world demonstrate the potential for this practice to contribute to environmental projects demonstrate the potential for this practice to contribute to environmental sustainability and economic development. For example:

1. **Malaysia:** The Malaysian Mangrove Honey Initiative engages local communities in sustainable beekeeping, combining mangrove restoration with honey production. The project educates locals on mangrove conservation and provides training in beekeeping skills, resulting in sustainable income generation.

2. **Thailand:** In coastal Thailand, beekeeping in mangrove areas has been promoted as part of ecotourism initiatives. Visitors participate in honey-harvesting activities, learning about the mangrove's importance and the value of sustainable beekeeping. The income from honey sales and tourism helps support community conservation efforts.

3. **The Sundarbans, India and Bangladesh:** The Sundarbans, the largest mangrove forest in the world, hosts traditional honey gatherers who have practiced sustainable honey harvesting for generations. Organizations are now working with these communities to improve beekeeping practices, enhance honey quality, and increase access to markets while promoting mangrove conservation.

These case studies illustrate that when beekeeping is aligned with conservation efforts and community empowerment, it can foster sustainable development and environmental resilience.

Future Prospects for Mangrove Beekeeping

The future of beekeeping in mangrove habitats looks promising, with increasing awareness of the value of mangrove honey and the need for sustainable environmental practices. However, scaling up this practice will require overcoming challenges related to environmental degradation, climate change, and the need for consistent training and education.

Greater investment in research and development will be crucial to understanding the specific needs of mangrove ecosystems and optimizing beekeeping practices in these habitats. For example, studies on the best mangrove species for nectar production, the impacts of climate change on flowering cycles, and the genetic diversity of local bee populations can help improve honey yields and colony health.

Furthermore, policy support from governments and non-governmental organizations (NGOs) will be needed to protect mangrove forests, regulate sustainable honey production, and facilitate access to markets for mangrove honey. Creating incentives for conservation, such as blue carbon credits for mangrove restoration projects, can encourage more communities to engage in sustainable beekeeping practices.

Beekeeping in mangrove habitats offers a unique opportunity to combine sustainable livelihoods with environmental conservation. By harnessing the ecological and economic benefits of mangrove honey production, communities can generate income while contributing to the health and resilience of coastal ecosystems. The practice supports biodiversity, stabilizes shorelines, and provides a natural solution for climate change mitigation through pollination and habitat protection.

As interest in sustainable and organic products continues to grow, mangrove honey has the potential to capture niche markets while promoting ethical and environmentally friendly practices. The path forward will involve addressing the challenges posed by environmental degradation, climate change, and market access, but with the right strategies and collaborations, beekeeping in mangrove habitats can thrive as a sustainable and impactful endeavor.

The Science Behind Mangrove Honey Production

Mangrove honey, a unique and often sought-after product, is produced in some of the world's most diverse and ecologically significant ecosystems. The relationship between mangrove trees, bees, and the environmental conditions that facilitate honey production is a fascinating area of study that combines aspects of botany, entomology, and environmental science. This article will explore the science behind mangrove honey production, detailing the specific mechanisms involved, the unique characteristics of mangrove honey, and the broader ecological implications of this process.

Understanding Mangrove Ecosystems

Before delving into the specifics of honey production, it's essential to understand what mangrove ecosystems are and why they are vital. Mangroves are coastal forests found in intertidal zones where saltwater and freshwater mix, usually in tropical and subtropical regions. These ecosystems are characterized by salt-tolerant trees and shrubs that have adapted to survive in harsh conditions, such as high salinity, tidal fluctuations, and anaerobic soils.

Mangrove forests play a critical role in coastal protection, biodiversity support, and climate regulation. They act as natural barriers against storms and erosion, provide habitats for a plethora of wildlife, and sequester carbon, contributing to climate change mitigation. The unique environmental conditions of mangroves also create a specialized habitat for bees, particularly for species that produce mangrove honey.

The Flora of Mangrove Ecosystems

Mangrove forests consist of various species, each with distinct ecological roles. The primary mangrove species include red mangroves (Rhizophora mangle), black mangroves (Avicennia germinans), and white mangroves (Laguncularia racemosa). These trees produce flowers that are rich in nectar, making them an excellent source of food for honeybees.

The floral composition of mangrove ecosystems is diverse and seasonally variable, which is crucial for the continuous production of honey. Different mangrove species bloom at different times of the year, ensuring that bees have a steady supply of nectar throughout the seasons. This flowering pattern is critical for sustaining bee colonies, as it provides the necessary resources for honey production.

The Role of Bees in Mangrove Honey Production

Bees are fascinating creatures that play an essential role in the production of mangrove honey. The primary species involved in this process is the honeybee (*Apis mellifera*), known for its efficient foraging behavior and social structure. Honeybees are attracted to the nectar produced by flowering mangrove trees, collecting it and transporting it back to their hives.

Upon returning to the hive, bees transform the nectar into honey through a process involving enzymatic activity and evaporation. The enzymes introduced during nectar collection break down the sugars in nectar into simpler sugars, which are more stable and less prone to crystallization. This process also reduces the water content of the nectar, allowing it to be stored as honey.

The unique environmental conditions of mangrove habitats influence the foraging behavior of bees. The brackish waters and saline air can affect the nectar's composition, leading to distinctive flavors in the resulting honey. The diversity of flowering species within mangrove forests also contributes to the complex flavor profiles of mangrove honey, making it a sought-after product.

Nectar Composition and Its Impact on Honey Quality

The quality of honey produced in mangrove ecosystems is significantly influenced by the nectar composition derived from mangrove flowers. Mangrove nectar contains a mixture of sugars, amino acids, vitamins, and minerals, which can vary based on several factors, including the species of mangrove tree, environmental conditions, and the time of year.

The primary sugars found in mangrove nectar are fructose and glucose, which are easily digestible and contribute to the sweetness of honey. The presence of various amino acids and micronutrients in the nectar enhances the nutritional profile of the honey, giving it potential health benefits. Some studies suggest that mangrove honey has higher antioxidant properties than honey produced from other floral sources, making it particularly valuable in traditional medicine and wellness practices.

Honey Production Process in Mangroves

The production of mangrove honey involves several stages, from nectar collection to the final product. Understanding this process is crucial for beekeepers aiming to maximize honey yields and ensure the quality of their product.

1. **Nectar Collection:**

   Honeybees begin by foraging for nectar during the blooming season of mangrove trees. Using their long proboscis, they extract nectar from the flowers and store it in their honey stomachs, a specialized organ designed for transporting nectar back to the hive.

2. **Enzymatic Activity:**

   Once the bees return to the hive, they pass the nectar to other worker bees through a process called trophallaxis, where the nectar is transferred mouth-to-mouth. During this transfer, enzymes such as invertase are introduced, breaking down the complex sugars in nectar into simpler sugars.

3. Evaporation:

After the enzymatic transformation, the nectar is spread across the honeycomb cells within the hive. Bees fan their wings to create airflow, facilitating the evaporation of excess water from the nectar. This evaporation process is critical for reducing moisture content and concentrating the sugars, ultimately converting nectar into honey.

4. Capping:

Once the honey reaches the desired moisture content (typically around 18%), bees seal the honeycomb cells with beeswax, preserving the honey for future use. Capping prevents moisture from re-entering the honey and protects it from spoilage.

Factors Influencing Honey Yield and Quality

Several environmental and biological factors influence the yield and quality of mangrove honey. Understanding these factors is essential for beekeepers looking to optimize their production:

1. Floral Diversity:

The diversity of mangrove species in a given area directly impacts the nectar availability and, consequently, honey yield. A diverse array of flowering mangrove trees can sustain bee populations throughout different seasons, ensuring a continuous supply of nectar.

2. Environmental Conditions:

Factors such as temperature, humidity, and rainfall play a significant role in flowering patterns and nectar production. For instance, periods of heavy rainfall may promote flowering in some mangrove species, while drought conditions can stress trees and limit nectar availability.

3. Bee Health and Management:

The health of bee colonies is critical for honey production. Healthy, well-managed colonies are more efficient foragers, leading to higher nectar collection rates and increased honey yields. Beekeepers must monitor their colonies for signs of disease, pests, or environmental stressors that could impact production.

4. Seasonal Variability:

Honey production is often seasonal, with peak yields occurring during specific flowering periods. Beekeepers must be aware of local flowering cycles and plan their management practices accordingly to maximize honey harvests.

The Unique Properties of Mangrove Honey

Mangrove honey possesses several unique properties that set it apart from honey produced in other environments. These properties are a result of the specific floral sources and environmental conditions associated with mangrove ecosystems.

1. **Flavor Profile:**

   Mangrove honey is known for its rich, complex flavor, which can vary depending on the dominant species flowering at the time of nectar collection. Common flavor notes include earthy, caramel-like, and slightly salty undertones, making it a delicacy among honey connoisseurs.

2. **Color and Texture:**

   The color of mangrove honey can range from dark amber to almost black, depending on the floral sources and processing methods. The texture is typically smooth and thick, with a luscious mouthfeel that appeals to consumers.

3. **Nutritional Benefits:**

   Studies have indicated that mangrove honey may contain higher levels of antioxidants compared to honey from other sources. These antioxidants contribute to its potential health benefits, which include anti-inflammatory and antimicrobial properties.

4. **Medicinal Uses:**

   In traditional medicine, mangrove honey is often used for its purported healing properties. It is believed to aid in wound healing, support digestive health, and provide relief for respiratory ailments. The unique compounds present in mangrove honey may play a role in its medicinal effectiveness.

Environmental and Economic Implications of Mangrove Honey Production

The production of mangrove honey has both environmental and economic implications that highlight its significance beyond mere consumption .

1. **Ecosystem Conservation:**

   By promoting sustainable beekeeping practices within mangrove ecosystems, honey production can serve as a catalyst for conservation efforts. Healthy bee populations contribute to the pollination and regeneration of mangrove forests, enhancing biodiversity and ecosystem resilience.

2. **Community Livelihoods:**

Mangrove honey production provides economic opportunities for local communities. By engaging in sustainable beekeeping practices, community members can generate income, improving their livelihoods while fostering a sense of stewardship for their natural resources.

3. **Ecotourism Potential:**

The unique appeal of mangrove honey can attract tourists interested in ecotourism and sustainable products. Beekeeping operations can serve as educational sites, allowing visitors to learn about the importance of mangrove ecosystems, honey production, and sustainable practices.

4. **Research and Education:**

The science behind mangrove honey production presents opportunities for research and education. By studying the interactions between bees and mangrove ecosystems, scientists can gain insights into pollination ecology, conservation strategies, and the impacts of climate change on these vital habitats.

Challenges Facing Mangrove Honey Production

Despite the potential benefits of mangrove honey production, several challenges must be addressed to ensure its sustainability:

1. **Environmental Degradation:**

Mangrove ecosystems face significant threats from coastal development, pollution, and climate change. These factors can lead to habitat loss, reduced nectar availability, and declining bee populations, jeopardizing honey production.

2. **Climate Change:**

Rising sea levels, increased temperatures, and altered precipitation patterns may disrupt the flowering cycles of mangrove trees, impacting nectar availability and honey production. Beekeepers must adapt to these changes by implementing management practices that account for shifting environmental conditions.

3. **Pest and Disease Management**

The health of bee colonies is critical for honey production. Beekeepers must monitor their hives for pests and diseases, such as Varroa mites and American foulbrood, which can severely impact bee populations.

4. Market Access and Education

Ensuring that local communities have access to markets for their mangrove honey is essential for sustainable production. Education and training programs can help beekeepers improve their practices, enhance honey quality, and connect with consumers.

Mangrove honey production is a remarkable interplay of ecological and economic factors that underscores the importance of sustainable practices in coastal ecosystems. The unique relationship between bees and mangrove trees not only contributes to the production of a valuable product but also supports biodiversity, promotes environmental conservation, and enhances community livelihoods. As awareness of the significance of mangrove ecosystems grows, so too does the potential for mangrove honey to play a vital role in both ecological and economic sustainability.

By investing in research, education, and sustainable management practices, we can ensure the continued production of mangrove honey and the preservation of the vital ecosystems from which it originates. The science behind mangrove honey production offers insights into the intricate connections that sustain our environment and highlights the need for continued efforts to protect and conserve these essential habitats for future generations.

Economic Benefits for Coastal Communities

Coastal communities, situated along the world's shorelines, have unique opportunities to harness natural resources and ecosystems for economic growth. These areas are home to diverse industries, ranging from fishing and tourism to renewable energy and aquaculture, each offering distinct pathways for local economic development. However, balancing economic activity with environmental sustainability is crucial to ensure long-term prosperity. This article explores various economic benefits that coastal communities can derive from their geographical and ecological resources, addressing key sectors, challenges, and strategies for sustainable development.

1. **Tourism and Recreation: A Major Economic Driver**

One of the most significant economic benefits for coastal communities is tourism. With their natural beauty, diverse wildlife, and cultural heritage, coastal regions attract millions of visitors every year, making tourism a vital source of income and employment.

- **Beaches and Marine Attractions:** Sandy beaches, coral reefs, and marine wildlife draw tourists who engage in recreational activities such as swimming, snorkeling, diving, and fishing. These activities generate revenue for local businesses, including hotels, restaurants, tour operators, and equipment rental services.

- **Cultural and Historical Sites:** Many coastal areas are rich in cultural heritage, with historic sites, maritime museums, and festivals that celebrate local traditions. Promoting cultural tourism adds economic value while preserving the cultural identity of the community.

- **Ecotourism and Sustainable Practices:** With a growing trend toward ecotourism, coastal communities can benefit from offering environmentally responsible travel experiences that focus on nature conservation. This approach not only generates income but also raises awareness about the importance of preserving coastal ecosystems.

2. **Fisheries and Aquaculture: Vital for Livelihoods**

Fisheries have historically been the backbone of many coastal economies, providing employment, food security, and export revenues. As fish stocks fluctuate due to overfishing and environmental changes, sustainable fisheries management and aquaculture have become essential.

- **Wild-Capture Fisheries:** Coastal communities rely on traditional fishing practices for local consumption and export markets. Adopting sustainable fishing techniques, such as catch limits and seasonal closures, can help ensure the long-term viability of fish stocks and support livelihoods.

- **Aquaculture Development:** Aquaculture, the farming of aquatic species like fish, shellfish, and seaweed, offers a promising alternative to wild-capture fisheries. It can provide stable income and employment opportunities while reducing pressure on natural fish populations. However, responsible management is required to minimize environmental impacts, such as water pollution and habitat destruction.

- **Value-Added Products:** Processing and marketing seafood products locally can increase profitability for coastal communities. Creating value-added products, such as smoked fish, canned seafood, or specialty shellfish, allows communities to capture a greater share of the seafood value chain.

3. **Renewable Energy: Harnessing Coastal Resources**

The push for renewable energy has created opportunities for coastal communities to tap into natural resources, such as wind, waves, and tides, to generate sustainable energy.

- **Offshore Wind Farms:** Coastal areas with strong and consistent winds are ideal locations for offshore wind energy development. These projects can bring significant economic benefits, including job creation in construction, maintenance, and operations, as well as lease revenues for local governments.

- **Wave and Tidal Energy:** Wave and tidal energy projects are emerging technologies with potential to provide reliable, renewable energy. By investing in these sectors, coastal communities can diversify their energy sources and reduce reliance on fossil fuels, contributing to energy security and sustainability.

- **Community-Based Energy Initiatives:** Community-owned renewable energy projects can help ensure that economic benefits are distributed locally. Revenue generated from these projects can be reinvested in local infrastructure, education, or social programs, fostering long-term economic resilience.

4. **Coastal Agriculture and Blue Economy Initiatives**

The concept of the "blue economy" encompasses sustainable use of ocean resources for economic growth, improved livelihoods, and ecosystem health. Coastal agriculture and other innovative blue economy initiatives present opportunities for sustainable development.

- **Seaweed and Algae Farming:** Coastal regions are ideal for cultivating seaweed and algae, which can be used for food, cosmetics, pharmaceuticals, and biofuels. Seaweed farming is environmentally friendly, as it absorbs carbon dioxide and provides habitat for marine life, while offering economic benefits through job creation and new market opportunities.

- **Salt Production and Desalination:** In some coastal areas, salt production remains a traditional livelihood. Modern desalination techniques can also provide freshwater resources for communities facing water scarcity. Developing technologies for sustainable desalination can enhance the economic resilience of coastal regions.

- **Marine Biotechnology:** The oceans are a source of diverse bioactive compounds with applications in medicine, agriculture, and industry. Investing in marine biotechnology research can stimulate economic growth by fostering innovation in pharmaceuticals, biofuels, and food production.

5. **Ports, Shipping, and Trade: Economic Gateways**

Ports play a crucial role in the economic development of coastal communities by facilitating trade, transportation, and logistics. Coastal regions with well-developed port infrastructure can benefit from trade-related activities that contribute to local economies.

- **Port Development and Expansion:** Expanding port facilities can increase trade volume, attract investment, and create jobs in shipping, warehousing, and logistics. Upgrading infrastructure to accommodate larger vessels and advanced technologies ensures the long-term competitiveness of the port.

- **Cruise Industry:** In addition to cargo shipping, the cruise industry provides economic benefits through tourism, as passengers disembark to explore coastal destinations. However, managing the environmental impact of cruise ships is necessary to protect the health of marine ecosystems.

- **Maritime Services and Shipbuilding:** Supporting maritime industries such as shipbuilding, repair, and maintenance can stimulate economic growth. Developing specialized maritime services, such as pilotage, navigation, and maritime law, adds further economic value.

6. **Coastal Protection and Ecosystem Services**

Healthy coastal ecosystems, such as mangroves, coral reefs, and wetlands, provide essential services that support the economic well-being of coastal communities.

- **Erosion Control and Storm Protection:** Natural barriers like mangroves and coral reefs reduce the impact of storms, preventing coastal erosion and protecting infrastructure. Investing in ecosystem restoration and conservation enhances resilience to climate change and natural disasters.

- **Carbon Sequestration:** Coastal ecosystems act as carbon sinks, absorbing carbon dioxide from the atmosphere. By protecting and restoring these habitats, coastal communities can contribute to global climate goals while benefiting from carbon credits or financial incentives for conservation efforts.

- **Biodiversity and Research Opportunities:** Coastal ecosystems are hotspots of biodiversity. Promoting research in marine science, ecology, and conservation can lead to new discoveries that support economic development in biotechnology, tourism, and education.

7. **Policy Frameworks and Community Engagement for Sustainable Development**

Realizing the full economic potential of coastal communities requires supportive policy frameworks and active community engagement.

- **Integrated Coastal Zone Management (ICZM):** ICZM is an approach that balances economic development with environmental protection. It involves stakeholder collaboration to manage resources sustainably, minimize conflicts, and optimize land and water use.

- **Community-Based Resource Management:** Empowering local communities to manage their resources fosters stewardship and ensures that economic benefits are equitably distributed. Community-led initiatives in fisheries, tourism, and renewable energy can drive sustainable development while respecting local traditions and knowledge.

- **Climate Change Adaptation and Resilience:** Coastal communities are particularly vulnerable to climate change impacts, such as sea-level rise, storms, and changing ocean conditions. Implementing adaptation strategies, including infrastructure upgrades, early warning systems, and ecosystem-based approaches, helps mitigate risks and secure economic stability.

Challenges to Economic Development in Coastal Areas

While coastal communities have numerous economic opportunities, they also face challenges that must be addressed to achieve sustainable growth:

1. **Environmental Degradation:** Coastal development, pollution, and resource exploitation can degrade ecosystems, reducing their ability to provide economic benefits. Sustainable practices and conservation efforts are essential to maintain the health of coastal environments.

2. **Climate Change Impacts:** Rising sea levels, ocean acidification, and extreme weather events pose significant risks to coastal economies. Strategies to build climate resilience are necessary to safeguard livelihoods and infrastructure.

3. **Regulatory and Governance Issues:** Complex regulatory frameworks and jurisdictional conflicts can hinder effective management of coastal resources. Streamlined governance and stakeholder collaboration are needed to resolve conflicts and promote integrated development.

4. **Economic Inequality:** Ensuring that the economic benefits of coastal development reach all community members is a challenge. Inclusive policies and initiatives that prioritize local employment, fair wages, and access to education can help address inequalities.

Coastal communities have a wealth of opportunities for economic development, provided they adopt sustainable practices that balance economic growth with environmental stewardship. From tourism and fisheries to renewable energy and the blue economy, coastal regions can derive significant economic benefits by harnessing their natural resources. However, addressing challenges such as environmental degradation, climate change, and regulatory issues is essential for long-term prosperity .

By fostering community engagement, implementing supportive policies, and prioritizing resilience, coastal communities can achieve sustainable economic development that benefits current and future generations. The path to economic success lies in recognizing the value of coastal ecosystems and investing in their preservation while capitalizing on their economic potential.

Case Studies: Successful Mangrove Beekeeping Projects

Mangrove beekeeping has emerged as a sustainable practice that brings ecological and economic benefits to coastal communities. It leverages the unique attributes of mangrove ecosystems, such as diverse flora and stable environmental conditions, to produce high-quality honey while supporting the preservation of these vital habitats. Across different regions, successful mangrove beekeeping projects have illustrated how community involvement, innovative techniques, and conservation efforts can transform local economies and bolster environmental resilience. This article highlights several case studies of mangrove beekeeping initiatives, providing insights into their strategies, achievements, and challenges.

1. **Kenya: Promoting Sustainable Livelihoods Through Mangrove Beekeeping**

In Kenya, coastal communities have long relied on mangroves for wood, fishing, and other resources, leading to overexploitation and degradation of these ecosystems. Recognizing the need for sustainable alternatives, beekeeping has been introduced as a means to generate income while reducing pressure on mangrove forests.

- **Project Overview:** In the coastal region of Kwale, several community-based organizations, supported by NGOs and government programs, began promoting mangrove beekeeping as a source of livelihood. The initiative involved training local communities in beekeeping techniques, providing equipment such as beehives and protective gear, and establishing honey processing facilities.

- **Outcomes:** The project has successfully increased household incomes through the sale of mangrove honey, with some beekeepers earning up to 30% more than before. Additionally, local participation in conservation efforts has grown, as community members understand the benefits of preserving mangroves for honey production.

- **Challenges and Solutions:** One significant challenge was the lack of knowledge about modern beekeeping techniques among local communities. To address this, training programs were conducted to teach sustainable practices such as hive management and pest control. Furthermore, efforts were made to establish market linkages to ensure a consistent demand for the honey produced.

2. **Indonesia: Integrating Mangrove Restoration with Beekeeping**

Indonesia, with its extensive mangrove forests, has been at the forefront of integrating beekeeping into mangrove restoration projects. The country's mangroves provide vital ecosystem services, including coastal protection and fishery support, but are under threat from deforestation and conversion for agriculture and aquaculture.

- **Project Overview:** In West Java, a project was launched to restore degraded mangrove areas by planting mangrove saplings and simultaneously establishing beekeeping activities. The initiative was a collaborative effort between local government agencies, universities, and community groups, aiming to create a sustainable livelihood model that promotes ecosystem restoration.

- **Outcomes:** The project successfully restored over 200 hectares of mangroves and established around 500 beehives managed by local beekeepers. The mangrove honey produced was marketed as a specialty product, fetching premium prices due to its unique flavor profile. Additionally, the restoration efforts improved fish populations, benefiting local fisheries.

- **Challenges and Solutions:** Restoration projects often face difficulties such as poor sapling survival rates and community reluctance to participate. In this case, local engagement was encouraged by providing financial incentives tied to the successful establishment of beekeeping enterprises. Training programs in sustainable beekeeping and honey marketing were also crucial for overcoming initial resistance.

3. **The Philippines: Empowering Women Through Mangrove Beekeeping**

In the Philippines, mangrove beekeeping has not only contributed to environmental conservation but also empowered women in coastal communities by providing them with income-generating opportunities.

- **Project Overview:** In the province of Palawan, a women-led initiative, supported by local NGOs, introduced mangrove beekeeping as a livelihood activity. The project aimed to provide economic empowerment while promoting mangrove conservation, with a focus on engaging women who often lacked access to income-generating activities.

- **Outcomes:** The initiative resulted in the establishment of multiple beekeeping groups managed by women, who were trained in hive management, honey processing, and marketing. The sale of mangrove honey and beeswax products has significantly improved the income of participants. Furthermore, the project contributed to a reduction in illegal mangrove cutting, as communities recognized the ecological and economic benefits of conservation.

- **Challenges and Solutions:** Gender norms in some communities posed challenges to women's participation in beekeeping. To address this, awareness-raising campaigns highlighted the role of women in environmental conservation and the economic advantages of beekeeping. Additionally, the project provided micro-loans to help women invest in beekeeping equipment, enabling them to start their enterprises.

4. **Vietnam: Community-Based Beekeeping and Ecotourism**

Vietnam's coastal communities have explored combining mangrove beekeeping with ecotourism to enhance local economies. The country's extensive mangrove forests, particularly in the Mekong Delta, offer unique opportunities for ecotourism development.

- **Project Overview:** A community-based initiative in the Ca Mau province sought to integrate mangrove beekeeping with ecotourism. Beekeepers were trained to manage hives within the mangrove forests, and tourists were invited to visit apiaries, learn about honey production, and participate in honey-tasting activities. The project aimed to generate income from both honey sales and ecotourism services.

- **Outcomes:** The initiative led to increased revenue for beekeepers, with income derived from honey sales as well as fees from ecotourism activities. The project also helped raise awareness about the importance of mangrove conservation. The increased tourism activity provided additional benefits to local businesses, such as homestays and restaurants, creating a positive economic ripple effect.

- **Challenges and Solutions:** Balancing the demands of tourism with conservation efforts was a challenge. To ensure sustainable ecotourism practices, guidelines were established to limit visitor numbers and minimize disturbances to the environment. Training in sustainable tourism management was also provided to community members involved in the initiative.

5. Thailand: Mangrove Beekeeping and Corporate Social Responsibility (CSR)

In Thailand, private sector involvement through Corporate Social Responsibility (CSR) programs has played a key role in supporting mangrove beekeeping projects.

- **Project Overview:** A multinational corporation partnered with local communities in the Chumphon province to promote mangrove conservation through beekeeping. The company provided funding for beekeeping equipment, training programs, and marketing support for mangrove honey. The project aimed to fulfill the company's CSR commitments while benefiting local livelihoods.

- **Outcomes:** The project facilitated the establishment of over 300 beehives and generated a steady income for participating households. The partnership also enhanced market access for the mangrove honey, with the product being sold both domestically and internationally. Furthermore, the company's involvement helped raise the profile of mangrove conservation efforts in the region.

- **Challenges and Solutions:** A challenge in this project was ensuring that community members had a genuine stake in the initiative and did not view it solely as a corporate exercise. The project addressed this by fostering local ownership of the beekeeping activities and establishing agreements that reinvested a portion of honey sales revenues into community development and conservation projects.

Key Lessons from Case Studies

1. **Community Involvement is Essential:** Successful projects emphasized the importance of involving local communities in planning, decision-making, and

management. This approach fosters ownership, ensures sustainability, and aligns the project's goals with local needs.

2. **Training and Capacity Building:** Providing adequate training in beekeeping practices, honey processing, and marketing is crucial for project success. Technical support helps participants overcome challenges and improve productivity.

3. **Market Access and Value Addition:** Establishing market linkages and promoting value-added products, such as specialty honey or honey-derived cosmetics, increases the economic returns for beekeepers. Branding mangrove honey as a unique, high-quality product can attract premium prices.

4. **Combining Conservation with Economic Incentives:** Integrating beekeeping with other activities like ecotourism or mangrove restoration provides multiple revenue streams and enhances the ecological benefits of the projects.

5. **Addressing Gender and Social Barriers:** Ensuring that marginalized groups, such as women, have opportunities to participate in beekeeping can lead to broader social benefits, including gender equality and community empowerment.

Mangrove beekeeping projects around the world illustrate how sustainable livelihood initiatives can simultaneously support economic development and environmental conservation. These case studies demonstrate the diverse ways in which coastal communities have successfully integrated beekeeping into their local economies, highlighting the importance of community participation, training, and market development. By learning from these experiences and addressing challenges, future projects can achieve even greater success, contributing to the sustainability of both livelihoods and mangrove ecosystems.

Challenges and Solutions in Mangrove Beekeeping

Mangrove beekeeping, while offering significant environmental and economic benefits, presents unique challenges due to the complexities of coastal ecosystems and the specific requirements of honey production in these habitats. From environmental issues and market access difficulties to technical limitations and social barriers, beekeepers in mangrove regions must navigate a range of obstacles to achieve sustainable success. This article explores some of the most pressing challenges in mangrove beekeeping and offers practical solutions that have been implemented in various regions to overcome them.

1. **Environmental Challenges: Navigating the Coastal Ecosystem**

Mangrove habitats are dynamic environments subject to tidal fluctuations, salinity changes, and seasonal weather patterns. These factors can impact beekeeping activities in several ways:

- **Tidal Inundation and Flooding:** Beehives located in coastal areas are vulnerable to tidal flooding, especially during high tides and storm surges. Flooding can damage hives, drown bees, and disrupt honey production.

- **Salinity and Water Quality:** High salinity levels can stress mangrove trees, potentially reducing their nectar output and affecting honey yields. Additionally, water pollution from coastal development, aquaculture, or oil spills can degrade the habitat and harm bee populations.

- **Solution:** Beekeepers can address tidal flooding by elevating beehives on stilts or platforms to protect them from rising water levels. Selecting hive locations with adequate elevation and natural barriers can also help. To manage salinity and water quality issues, conservation measures such as mangrove restoration and pollution control initiatives are essential to maintain a healthy ecosystem that supports bee activity.

2. **Limited Floral Diversity: Seasonal Nectar Shortages**

Mangroves, while rich in nectar-producing species, may not provide a continuous supply of forage throughout the year. Seasonal variations in flowering can lead to periods of nectar scarcity, affecting bee nutrition and honey production.

- **Challenge:** During non-flowering seasons, bees may struggle to find sufficient nectar and pollen, resulting in lower honey yields and weakened colonies.

- **Solution:** Beekeepers can diversify their forage sources by introducing supplementary feeding during lean seasons. Providing sugar syrup, pollen substitutes, or protein supplements can help maintain colony strength. Additionally, planting a variety of native, nectar-producing plants around the mangrove area can extend the foraging period and offer a more consistent food supply for bees.

3. **Predation and Pests: Threats to Bee Health**

Mangrove regions are home to various predators and pests that pose risks to bee colonies. These include larger animals like birds and crabs, as well as smaller pests such as ants, mites, and beetles.

- **Challenge:** Predators can damage hives and eat the bees, while pests can introduce diseases, infest hives, and weaken or kill colonies. For example, the Varroa mite is a common and serious threat to bee health, causing colony collapse if not managed.

- **Solution:** Implementing hive management practices such as regular inspections, use of screened bottom boards, and applying natural pest deterrents like essential oils can help reduce pest infestations. Placing hives on platforms or stilts can also protect them from ground-based predators. For more severe infestations, beekeepers may need to resort to targeted treatments, using organic methods wherever possible to avoid harming the bees.

4. **Market Access and Value Addition: Overcoming Economic Barriers**

Despite the growing demand for mangrove honey, accessing markets can be challenging for beekeepers, particularly in remote coastal communities. Issues include a lack of market information, inadequate marketing channels, and competition from other types of honey.

- **Challenge:** Without proper market access and marketing strategies, beekeepers may struggle to sell their products at fair prices. Moreover, the lack of branding and differentiation can result in lower profit margins.

- **Solution:** Developing market linkages through cooperatives or associations can improve bargaining power and facilitate the sale of honey. Additionally, branding mangrove honey as a premium, eco-friendly product with unique characteristics (such as a distinct flavor profile) can help command higher prices. Beekeepers can also explore producing value-added products like beeswax candles, cosmetics, or medicinal honey to diversify income streams.

5. **Technical Challenges: Gaps in Beekeeping Knowledge and Equipment**

Many coastal communities where mangrove beekeeping is practiced have limited access to modern beekeeping equipment and techniques. This can hinder productivity and make it difficult to scale up operations.

- **Challenge:** Lack of knowledge about hive management, honey extraction, and quality control can result in suboptimal honey yields and poor product quality. Additionally, the high cost of specialized equipment, such as honey extractors, can be prohibitive for small-scale beekeepers.

- **Solution:** Training programs that teach best practices in beekeeping, hive management, and honey processing can significantly improve productivity and honey quality. Establishing shared processing facilities or cooperatives can help reduce individual costs associated with purchasing equipment. NGOs and government programs can also play a role by subsidizing equipment costs or providing financial assistance to beekeepers.

6. **Social and Cultural Barriers: Addressing Gender and Community Dynamics**

In some coastal communities, social norms and cultural practices may limit participation in beekeeping activities, especially for women. Additionally, community members may be hesitant to adopt new practices due to traditional beliefs or fear of bee stings.

- **Challenge:** Gender norms may restrict women's involvement in beekeeping, leading to a lack of opportunities for economic empowerment. Resistance to change and misconceptions about bees can also impede the adoption of beekeeping as a viable livelihood.

- **Solution:** Engaging community leaders and conducting awareness campaigns can help change perceptions and encourage broader participation in beekeeping. Creating women-led beekeeping groups or cooperatives has been successful in some regions, empowering women to take leadership roles and providing income-generation opportunities. Offering protective equipment and hands-on training can also alleviate fears and build confidence in working with bees.

7. **Climate Change Impacts: Adapting to a Changing Environment**

Climate change poses long-term challenges to mangrove beekeeping through its effects on weather patterns, sea-level rise, and mangrove health. Changes in rainfall, temperature, and storm frequency can disrupt flowering cycles and affect bee foraging behavior.

- **Challenge:** Unpredictable weather patterns can lead to irregular nectar flows, increased risk of hive damage, and higher bee mortality rates. Rising sea levels may also threaten the very existence of mangrove habitats.

- **Solution:** Implementing adaptive beekeeping practices, such as relocating hives to higher ground, establishing backup apiaries in more stable environments, and planting climate-resilient forage species can help mitigate the impacts of climate change. Supporting mangrove conservation and restoration initiatives will also be essential to preserving these ecosystems for future generations.

8. **Regulatory and Policy Challenges: Navigating Legal Frameworks**

In some countries, beekeeping regulations may not be well-developed, or the legal frameworks governing mangrove conservation and land use may present obstacles to beekeeping activities.

- **Challenge:** Complex permitting processes, restrictions on land use, and unclear guidelines on mangrove management can make it difficult for beekeepers to establish operations legally.

- **Solution:** Advocacy efforts to simplify regulations and promote beekeeping-friendly policies can help facilitate legal compliance. Engaging with government agencies to develop integrated management plans that recognize beekeeping as a sustainable livelihood option is also beneficial. Collaboration with NGOs and research institutions can support the development of policy recommendations based on successful case studies.

9. **Health and Safety Concerns: Ensuring Safe Practices**

Working with bees in a mangrove environment poses health and safety risks, including bee stings, allergic reactions, and the dangers associated with operating near water.

- **Challenge:** Inadequate safety measures can result in accidents or discourage people from participating in beekeeping activities.

- **Solution:** Providing proper protective gear, including beekeeping suits, gloves, and veils, can reduce the risk of stings. Training in safety procedures, such as hive handling techniques and first aid for bee stings, is also crucial. For apiaries located near water, implementing safety measures such as life vests and having emergency plans in place can further protect beekeepers.

While mangrove beekeeping presents a range of challenges, it also offers significant opportunities for sustainable development, environmental conservation, and community empowerment. The experiences of various mangrove beekeeping projects highlight the importance of adaptive management practices, community involvement, and multi-stakeholder collaboration in overcoming obstacles. By addressing the challenges and implementing the solutions outlined in this article, beekeepers can maximize the benefits of mangrove beekeeping and contribute to the long-term sustainability of coastal ecosystems.

Future Prospects for Mangrove Honey and Conservation

Mangrove honey, derived from the nectar of mangrove blossoms, is not just a unique natural product; it is a symbol of the vital relationship between coastal ecosystems and sustainable livelihoods. With its distinctive taste and health benefits, mangrove honey has gained recognition as a high-value commodity. Beyond its economic potential, the production of this honey is linked to crucial environmental benefits, including mangrove conservation and biodiversity support. As global awareness of environmental sustainability grows, the future prospects for mangrove honey and related conservation efforts appear promising. This article will explore the emerging trends, opportunities, and challenges for the growth of mangrove honey production and the conservation of mangrove ecosystems.

1. **Expanding Market Opportunities for Mangrove Honey**

The demand for natural and organic products is on the rise globally, driven by health-conscious consumers who seek high-quality, eco-friendly alternatives. Mangrove honey fits this trend, given its rich antioxidant content, unique flavor, and medicinal properties.

- **Premium Positioning and Niche Markets:** Mangrove honey can be positioned as a premium product due to its distinct characteristics. Its rarity and limited production create a niche market opportunity where consumers are willing to pay higher prices for high-quality, sustainably sourced honey. This positioning can be further enhanced by certification programs such as organic or fair-trade labeling, which boost consumer confidence.

- **Potential for Value-Added Products:** Beyond pure honey, there are opportunities to develop value-added products such as honey-infused skincare products, supplements, or gourmet food items. Expanding the product range can attract different market segments, increasing revenue potential for producers.

- **Export Potential:** With increasing recognition of the unique attributes of mangrove honey, there is potential to expand into international markets. Countries in the European Union, North America, and Asia, where the demand for natural and medicinal honey is growing, could serve as lucrative markets for mangrove honey exports.

2. **Leveraging Mangrove Honey for Conservation**

Mangrove beekeeping is not just about honey production; it is also a strategy for promoting mangrove conservation. Beekeeping offers coastal communities a sustainable alternative to destructive practices, such as logging and conversion of mangroves for aquaculture or agriculture.

- **Incentivizing Conservation Through Economic Benefits:** The income generated from mangrove honey production can serve as an incentive for local communities to actively participate in mangrove conservation. By providing an economic alternative, beekeeping encourages the preservation and restoration of mangrove forests, which serve as critical habitats for a wide variety of marine and terrestrial species.

- **Integrating Beekeeping with Mangrove Restoration Projects:** Many mangrove restoration initiatives now incorporate beekeeping as a component of ecosystem rehabilitation. This approach provides dual benefits—restoring degraded habitats and creating income-generating opportunities. As more projects adopt this integrated model, there will likely be a proliferation of mangrove honey production linked to large-scale restoration efforts.

- **Raising Awareness Through Community-Led Conservation:** Beekeeping programs that involve local communities play a crucial role in raising awareness about the ecological value of mangroves. When community members actively participate in conservation efforts and benefit economically from these ecosystems, they become natural stewards of the environment, driving grassroots conservation movements.

3. **Innovations in Beekeeping Techniques**

Technological advancements and innovative practices in beekeeping can significantly enhance mangrove honey production, making it more efficient and sustainable.

- **Modern Hive Designs:** Introducing beehives with features that improve honey yield and protect bees from environmental threats (e.g., flooding or predation) can enhance productivity. Elevated or waterproof hives are particularly useful in mangrove environments, where tidal fluctuations pose a risk.

- **Utilizing Data and Technology for Hive Monitoring:** The use of technology, such as smart hive monitoring systems, allows beekeepers to track hive conditions, including temperature, humidity, and hive weight. This data-driven approach can help in identifying potential issues, optimizing hive management, and improving honey yield.

- **Selective Breeding for Resilient Bee Species:** Developing or selecting bee species that are more resilient to local environmental conditions (e.g., high humidity, pests) can help increase hive survival rates and productivity. These resilient bee strains may also adapt better to the unique floral resources available in mangrove ecosystems.

4. **Addressing Challenges for Sustainable Growth**

While the future of mangrove honey production is promising, it is not without challenges. Addressing these obstacles is key to ensuring the long-term sustainability of both honey production and mangrove conservation.

- **Climate Change and Environmental Risks:** Mangroves are vulnerable to the impacts of climate change, such as sea-level rise, extreme weather events, and temperature fluctuations. These changes can affect flowering patterns, nectar availability, and the health of bee colonies. Adaptive management strategies, including relocating hives during extreme weather and integrating climate-resilient mangrove species, will be crucial.

- **Ensuring Quality and Consistency:** As the market for mangrove honey grows, ensuring consistent quality becomes a priority. Establishing standardized processing and quality control measures can help maintain the reputation of mangrove honey as a high-value product. Implementing certification schemes for sustainable production will also strengthen consumer trust.

- **Overcoming Regulatory Barriers:** In some countries, regulatory frameworks governing beekeeping, honey production, and land use in coastal areas are either underdeveloped or overly restrictive. Streamlining regulations and creating policies that support sustainable mangrove honey production could facilitate the growth of the industry.

5. **Policy Support and Multi-Stakeholder Collaboration**

The success of mangrove honey production and conservation efforts hinges on supportive policies and collaboration among various stakeholders, including governments, NGOs, research institutions, and the private sector.

- **Government Incentives for Sustainable Practices:** Governments can play a role in promoting sustainable mangrove honey production by providing financial incentives, such as subsidies for beekeeping equipment or tax breaks for conservation-friendly practices. Policies that integrate mangrove restoration into national environmental strategies can also help scale up efforts.

- **NGO and Community Partnerships:** Non-governmental organizations often spearhead conservation and sustainable livelihood initiatives in coastal areas. Partnerships with local communities are crucial for capacity-building and project implementation. Successful collaborations can offer valuable models for replication in other regions.

- **Private Sector Engagement:** The private sector can contribute to mangrove conservation through Corporate Social Responsibility (CSR) programs, investments in sustainable supply chains, and support for mangrove honey as a high-quality product. By partnering with producers, companies can help create market access, improve production standards, and promote sustainable sourcing.

6. **The Role of Research and Education**

Research and education play a vital role in advancing the prospects of mangrove honey and conservation by driving innovation, enhancing knowledge, and raising awareness.

- **Research on Honey Properties and Mangrove Ecosystems:** Scientific studies can help uncover additional health benefits of mangrove honey, adding to its value proposition. Research on the ecological dynamics of mangrove ecosystems and their relationship with beekeeping can also inform better management practices and policies.

- **Capacity Building Through Education Programs:** Training and education programs that teach sustainable beekeeping techniques, ecosystem management, and business skills are crucial for the success of mangrove honey initiatives. These programs empower communities to adopt best practices and improve the quality of their products.

- **Promoting Public Awareness:** Public awareness campaigns that highlight the connection between mangrove honey and conservation can help garner support for sustainable production and environmental protection. When consumers understand the broader impact of their purchasing decisions, they are more likely to support eco-friendly products.

7. Potential Environmental and Social Impacts

The future of mangrove honey and conservation extends beyond economic and ecological outcomes; it also involves social and cultural dimensions.

- **Empowering Coastal Communities:** Mangrove honey production offers an alternative livelihood for coastal communities, helping to reduce poverty and reliance on environmentally damaging activities. Empowering these communities can lead to broader social benefits, including improved health, education, and gender equality.

- **Supporting Biodiversity Conservation:** Beekeeping encourages the protection of mangrove forests, which are home to a diverse array of species. By promoting sustainable beekeeping practices, the conservation of these ecosystems can support broader biodiversity goals, including the protection of endangered species.

- **Strengthening Climate Resilience:** Mangroves play a key role in climate change mitigation and adaptation by sequestering carbon, stabilizing shorelines, and buffering against storms. Integrating beekeeping with conservation efforts can help strengthen these climate resilience benefits, providing a win-win scenario for both nature and society.

Mangrove honey represents a unique intersection of sustainable livelihoods and environmental conservation. As the world moves toward more sustainable consumption patterns, the future prospects for mangrove honey production and the conservation of mangrove ecosystems are bright. Expanding market opportunities, integrating conservation with economic incentives, adopting innovative techniques, and addressing challenges will be key to unlocking the full potential of this practice. With collaborative efforts from governments, communities, and the private sector, mangrove beekeeping can pave the way for a future where ecological health and economic prosperity go hand in hand.

Policy and Support for Mangrove Conservation and Beekeeping

Mangrove ecosystems are among the world's most productive and biologically diverse habitats, providing essential services such as carbon sequestration, coastal protection, and support for fisheries. Despite their importance, mangroves face significant threats from coastal development, deforestation, and climate change. Beekeeping in mangrove areas offers a sustainable way to support conservation while creating economic opportunities for local communities. Effective policies and support mechanisms are crucial for the success of mangrove conservation and beekeeping initiatives, ensuring the long-term sustainability of these ecosystems.

1. **The Importance of Mangrove Ecosystems**

Mangroves play a vital role in coastal and marine ecosystems. Their complex root systems stabilize shorelines, reduce erosion, and act as natural barriers against storm surges. Mangroves are also highly efficient at storing carbon, making them a key player in mitigating climate change. They serve as nurseries for various fish species and provide habitats for numerous wildlife, including birds, crabs, and insects.

However, mangrove forests continue to decline globally due to human activities. Policies that support mangrove conservation and sustainable livelihoods, such as beekeeping, are essential to protect these valuable ecosystems from further degradation.

2. **Policies Supporting Mangrove Conservation**

Governments and international organizations have developed various policies to promote mangrove conservation, ranging from regulatory frameworks to financial incentives:

- **Protected Area Designation:** Designating mangroves as protected areas can limit human activities that harm these ecosystems, such as logging, coastal development, and industrial waste discharge. Establishing marine protected areas (MPAs) that include mangrove forests helps ensure their long-term preservation.

- **Mangrove Restoration Programs:** National and regional programs aimed at restoring degraded mangrove forests are essential to reversing habitat loss. Policies that incentivize restoration through funding, technical assistance, or public-private partnerships encourage large-scale restoration efforts. Restoration projects often involve community participation, ensuring that local people benefit from conservation initiatives.

- **Land-Use Regulations:** Zoning laws and coastal management policies can be used to restrict activities that lead to the destruction of mangroves, such as the expansion of shrimp farms, hotels, or other development projects. Policies that integrate mangrove conservation into land-use planning help balance economic growth with environmental sustainability.

3. **Beekeeping as a Sustainable Livelihood**

Beekeeping in mangrove habitats offers an environmentally friendly way to generate income while supporting conservation. Mangrove honey, known for its distinct flavor and medicinal properties, is a valuable product that can be marketed as a premium commodity. Beekeeping helps incentivize the protection of mangrove ecosystems by providing communities with an alternative income source that does not involve the exploitation of the forest.

- **Income Generation for Coastal Communities:** Beekeeping enables coastal communities to diversify their livelihoods. It provides a stable income, especially in areas where fishing and agriculture are affected by environmental changes. By participating in sustainable honey production, communities gain a vested interest in protecting mangrove forests.

- **Pollination Services and Biodiversity Benefits:** Bees play a critical role in pollinating plants, which supports biodiversity and improves the resilience of mangrove ecosystems. Promoting beekeeping in mangrove areas can enhance the health of the entire ecosystem, benefiting other species that rely on mangroves for habitat.

4. **Policy Measures to Promote Mangrove Beekeeping**

To maximize the benefits of mangrove beekeeping for both conservation and community development, targeted policies and support mechanisms are needed:

- **Financial Incentives for Beekeepers:** Governments can introduce subsidies, grants, or low-interest loans to help beekeepers cover the costs of starting and maintaining their hives. Providing financial support for equipment and training can encourage more people to participate in beekeeping activities. Incentives can also be linked to sustainable practices, rewarding beekeepers who engage in conservation efforts.

- **Capacity Building and Training Programs:** Education and training initiatives are critical to improving beekeeping practices, honey quality, and business skills. Policies that support technical training for beekeepers, including workshops on hive management, honey extraction, and marketing, can enhance productivity and income potential. Training programs can also promote awareness of environmental issues, such as the importance of conserving mangrove habitats.

- **Certification and Labeling Schemes:** Implementing certification schemes for sustainable mangrove honey production can help differentiate this product in the market, allowing beekeepers to charge premium prices. Certification programs can include criteria for organic production, fair trade, or eco-friendly practices, helping consumers identify products that contribute to conservation.

5. Integrating Mangrove Beekeeping with National Environmental Strategies

National environmental strategies often include targets for biodiversity conservation, climate change mitigation, and sustainable development. Integrating mangrove beekeeping into these strategies can amplify the impact of conservation efforts.

- **National Biodiversity Action Plans:** Including mangrove beekeeping in national biodiversity strategies can promote the dual goals of species conservation and sustainable livelihood development. Beekeeping can be recognized as an important activity for maintaining the health of ecosystems, and policies can be created to support beekeepers in regions with high biodiversity.

- **Climate Change Mitigation and Adaptation:** Mangroves are significant carbon sinks, and their preservation contributes to climate change mitigation. Incorporating beekeeping into climate adaptation strategies can also enhance community resilience. Beekeeping generates income and provides alternative food sources, helping communities adapt to the impacts of climate change on traditional livelihoods such as fishing.

- **Partnerships with NGOs and International Organizations:** Collaborating with non-governmental organizations and international bodies can enhance policy effectiveness. NGOs often bring technical expertise, financial resources, and global networks that can be leveraged to support local beekeeping and conservation initiatives. International cooperation can also facilitate knowledge-sharing, helping countries learn from successful mangrove conservation programs worldwide.

6. Case Studies: Successful Policy Implementation

Examining successful case studies of policy implementation can offer insights into best practices for supporting mangrove conservation and beekeeping.

- **The Philippines:** The Philippine government has implemented several programs aimed at restoring mangrove forests and promoting sustainable livelihoods. Policies that incentivize community involvement in mangrove restoration have resulted in increased honey production and improved coastal resilience. The Department of Environment and Natural Resources has also collaborated with NGOs to provide training on sustainable beekeeping techniques, benefiting local economies.

- **Kenya:** In Kenya, the development of mangrove honey as a premium product has been supported by partnerships between local communities, NGOs, and the government. Certification programs for organic and fair trade honey have enabled beekeepers to access international markets, boosting incomes and encouraging the protection of mangrove forests.

- **Indonesia:** The Indonesian government has integrated mangrove conservation into its national climate change strategy, with beekeeping recognized as a viable livelihood alternative. Through government support and international funding, mangrove restoration projects have been paired with beekeeping initiatives, leading to increased community participation in conservation efforts.

7. Addressing Barriers to Effective Policy Implementation

While policies can significantly support mangrove conservation and beekeeping, there are also barriers to overcome.

- **Policy Gaps and Inconsistencies:** In some regions, policies governing land use, beekeeping, and conservation are fragmented or contradictory. Aligning policies across sectors is essential to ensure that mangrove conservation efforts are not undermined by competing interests, such as aquaculture expansion or tourism development.

- **Limited Access to Finance and Resources:** Many small-scale beekeepers face difficulties accessing credit or financial assistance to start and maintain their businesses. Policy interventions that facilitate access to finance, such as microcredit schemes or grants, can help bridge this gap.

- **Social and Cultural Challenges:** Beekeeping may not be culturally accepted or may face gender-related barriers in some communities. Policies should consider social dynamics and include measures to promote gender equality and community engagement in beekeeping activities.

Mangrove conservation and beekeeping present a powerful opportunity to achieve environmental and economic sustainability. Policies that support the sustainable use of mangrove resources, incentivize beekeeping, and integrate conservation into national strategies can create a framework for success. By addressing barriers to implementation and promoting inclusive and innovative approaches, the future of mangrove honey production and conservation looks bright. Collaborative efforts from governments, communities, NGOs, and international organizations will be essential in making this vision a reality.

Conclusion: The Sweet Path to Sustainability

The symbiosis between mangrove conservation and beekeeping offers a powerful vision for sustainable development that bridges economic and ecological interests. The cultivation of mangrove honey presents a unique opportunity to address the pressing issues facing coastal ecosystems while providing meaningful livelihoods for local communities. As environmental and economic challenges mount, finding innovative, nature-based solutions like this one becomes essential for creating a sustainable future.

Mangroves play an indispensable role in coastal protection, biodiversity, and climate regulation. Their ability to sequester carbon, buffer against storm surges, and support a diverse range of marine and terrestrial life makes them one of the most valuable ecosystems on the planet. However, they are under constant threat from human activities such as deforestation, coastal development, and aquaculture. The decline of mangrove forests represents a significant loss, not only for the environment but also for the communities that depend on them. This is where mangrove honey production comes into play—offering a sustainable alternative that aligns community livelihoods with conservation goals.

Beekeeping in mangrove areas serves as an incentive for communities to protect these forests. By generating income through the sale of high-value mangrove honey, coastal populations are provided with a direct economic reason to preserve and restore mangroves. The practice discourages harmful activities, such as logging or converting mangroves for shrimp farming, which have long-term ecological costs. Instead, it encourages a culture of stewardship, where the well-being of the ecosystem directly impacts the community's economic prospects.

The market potential for mangrove honey is vast, driven by growing consumer demand for natural and ethically sourced products. Its unique flavor and medicinal qualities, coupled with the ecological narrative behind its production, create a compelling value proposition. Positioning mangrove honey as a premium product opens opportunities for niche markets, including organic, fair-trade, and artisanal food segments. Value-added products, such as honey-based skincare items or health supplements, further extend its commercial potential, making it a versatile income source.

As awareness about sustainable consumption rises, consumers are increasingly seeking products that contribute positively to the environment. Mangrove honey is more than a commodity; it is a symbol of conservation, a testament to the harmonious relationship between people and nature. This appeal can be leveraged to promote broader environmental goals, encouraging consumers to support products that benefit ecosystems and the communities that protect them. Certification schemes, such as fair-trade labeling or organic standards, can also help differentiate mangrove honey in the market, assuring consumers that their purchases contribute to sustainable practices.

However, the sweet path to sustainability is not without its challenges. Climate change poses significant risks to mangrove ecosystems and the practice of beekeeping within them. Rising sea levels, extreme weather events, and changes in temperature can affect the health of both the mangroves and the bee populations that depend on them. Policies must therefore account for adaptive management strategies that enhance resilience, such as planting climate-resistant mangrove species, using elevated hive designs to protect against flooding, and integrating beekeeping into broader climate adaptation plans.

Moreover, to maximize the potential of mangrove honey production, supportive policies must extend beyond conservation. They should also encompass financial assistance, capacity building, and market development. Grants, subsidies, or microfinance programs can help small-scale beekeepers overcome initial costs, while training initiatives can improve skills in hive management, honey processing, and business development. Establishing accessible markets for honey products, both locally and internationally, is crucial for enabling producers to capitalize on the economic benefits of mangrove beekeeping.

Integrating mangrove honey production into national and regional conservation strategies requires collaborative efforts from governments, NGOs, research institutions, and the private sector. Policymakers should create enabling environments where sustainable practices are not only encouraged but rewarded. This includes aligning regulatory frameworks to facilitate mangrove restoration projects and ensuring that beekeeping is recognized as a valuable livelihood within coastal management plans. NGOs can play a role in capacity building and project implementation, while private sector partnerships can provide market access and investment.

Education and awareness campaigns are also essential components of this strategy. Teaching communities about the ecological importance of mangroves and the economic potential of sustainable beekeeping can foster a culture of conservation. When communities see the tangible benefits of protecting their natural resources, they are more likely to engage in conservation efforts. Schools, local organizations, and public outreach programs can all contribute to this educational mission, making the connection between the health of the ecosystem and the prosperity of the community clear.

The practice of mangrove beekeeping also extends its benefits beyond economics and conservation. Socially, it empowers communities, providing alternative livelihoods that can help reduce poverty and improve social conditions. For marginalized groups, such as women and indigenous communities, beekeeping offers an accessible and sustainable source of income, potentially leading to greater financial independence and social inclusion. Programs that focus on gender equality in beekeeping and involve women in training initiatives can significantly amplify these benefits.

From a biodiversity perspective, beekeeping supports the pollination of mangrove trees and other flora, enhancing the resilience and productivity of these ecosystems. Healthy mangroves, in turn, support a wider array of marine and terrestrial life, contributing to overall ecosystem health. The interdependent relationship between bees and mangroves creates a positive feedback loop where the preservation of one supports the thriving of the other, demonstrating the interconnectedness of species and ecosystems.

In conclusion, the path to sustainability through mangrove honey production represents a multifaceted approach that addresses ecological, economic, and social dimensions. It is a pathway that encourages not just the conservation of mangrove forests but also the empowerment of the people who rely on them. As more stakeholders come together to support this endeavor, the potential for a scalable and impactful model of sustainable development becomes increasingly attainable. The sweet path to sustainability is about creating a future where thriving communities and healthy ecosystems coexist, driven by practices that respect nature and promote long-term well-being.

By embracing mangrove honey production as part of a broader conservation strategy, society can take meaningful steps toward addressing some of the most pressing challenges of our time. It is not just about preserving forests or selling honey; it is about shaping a world where human prosperity is intrinsically linked to the health of the planet. Through the combined efforts of communities, policymakers, and consumers, we can indeed forge a sustainable future—one drop of honey at a time.

www.ingramcontent.com/pod-product-compliance
Lightning Source LLC
Chambersburg PA
CBHW062126220526
45471CB00010B/3900